SIMPLY FABRIC

50 CREATIVE IDEAS FOR IMPROVING YOUR HOME

LINDA BARKER

PHOTOGRAPHY BY LIZZIE ORME

SIMPLY
FABRIC

50 CREATIVE IDEAS FOR IMPROVING YOUR HOME

LINDA BARKER
PHOTOGRAPHY BY LIZZIE ORME

ANAYA PUBLISHERS LTD
LONDON

For Ro

First published in Great Britain in 1993 by

Anaya Publishers Ltd
Strode House, 44-50 Osnaburgh Street, London NW1 3ND

Copyright © Anaya Publishers Ltd 1993

Series Editor	Janet Donin
Designer	Jerry Goldie
Photographer	Lizzie Orme
Assisted by	Sussie Nielsen
Stylist	Linda Barker
Assisted by	Sharon Williams
	Diane Crawford

British Library Cataloguing in Publication Data
Linda Barker
Simply Fabric. – (Simply Series)
I. Title   II. Series
747
ISBN 1-85470-083-9
Typeset in Great Britain by Bookworm Typesetting, Manchester
Colour reproduction by Scantrans Pte, Singapore
Printed and bound in Italy by Officine de Agostini Spa, Novarra

Contents

Introduction
7

1
No Sew
13

2
Easy Sew
35

3
Quick Stitch
63

4
All Stitched Up
87

INTRODUCTION

The look and texture of fabric can work wonders in your home, so why not try some of my simple projects?

I don't know about you but sometimes I could happily throw my sewing machine out of the window. So don't think that just because this is a book on fabric, it's not useful to those of you who would rather sleep on a bed of nails than attempt to transform ten metres of fabric into a pair of curtains. And don't you just hate those people who can 'run up' a pair of fully lined, swagged, piped and trimmed curtains in an afternoon? And all in between doing the dishes and making dinner! Well you and I both! So rest assured that in SIMPLY FABRIC there are projects for sewers of varying standards and for those with a passionate dislike for all things that require stitching. This book is not about how to attach silken linings to flamboyant, pinch-pleated dress curtains; or how to re-upholster your granny's favourite armchair. There are plenty of technical books on the market that, quite honestly, can provide more of that kind of information than I am able to do here. However, this is a book about style and a love of the wonderful colours and textures that fabric can bring into your home. I'll show you how to use the barest minimum of an expensive fabric to get maximum good looks – without spending a fortune! But equally I'll show you how to salvage an unappealing fabric, simply by cutting it up and re-using it with a decoupage technique.

Many of the projects in this book may inspire you to make your own interpretations, so don't think you have to follow all my ideas slavishly. If you like the look of the folding screen, but don't really need one in the bedroom, then add three or more panels and you have a clever room divider for the living/dining room. Similarly, if you love the effect of the leafy picture frame, but you don't have a suitable picture, then simply stick the leaves around your bathroom mirror, or use the same technique to give a pretty edge to a wicker basket or bowl of pot pourri. When you discover that working with fabric isn't all frills and flounces, you may be persuaded to sew a few of the simpler projects. Let's face it, you won't be able to find my wonderful kitchen blinds, made out of old-fashioned tea cloths, anywhere – no matter where you look! Which is why SIMPLY FABRIC shows you how to have fun while still being creative. And with my fifty step-by-step projects I know you'll find plenty of tempting ideas to work in your home.

For those of you without a sewing machine, you'll find that many of the projects can easily be sewn by hand, especially those in the No Sew and Easy Sew chapters. You may also be inspired to attempt the simple ticking cushions and bedroom cushions in the Quick Stitch chapter.

Here too, you will find simple, practical approaches to working with fabric, as well as some more flirtatious examples. For instance, I hope that you will be pleasantly surprised at how the well-known, yet sadly over-used festoon blind, can look refreshingly different when given a new application as a decorative curtain heading.

I hope that by using SIMPLY FABRIC you will gain the confidence to try these projects as well as some new ideas of your own.

EQUIPMENT

Much as I would love it, I am not the owner of a super, highly technical sewing machine, so you shouldn't feel that without this sort of complicated hardware you're not going to be able to tackle the ideas in this book. In fact, I'm quite certain that even with an ancient, hand-cranked, manual machine you will be able to achieve most of the makes. It's worthwhile remembering that curtains were hung at windows long before the introduction of the sewing machine! I acquired my second-hand machine over ten years ago and despite the occasional urge to throw it out of the nearest window, it is still functioning perfectly. If you do decide to buy a new machine make sure you test it well before you buy and if possible arrange for a home trial.

All you need to create most projects in SIMPLY FABRIC, is the basic running stitch. A zigzag function which prevents hems and cut edges from fraying, is also useful. However as an alternative, you can use a running stitch along an edge and then use pinking scissors to stop the edges from becoming ragged.

Sewing box After a while you will find that it makes good sense to keep all your sewing paraphernalia together. There are many sewing baskets available, but I find the best storage box around is one of those plastic tool boxes from a do-it-yourself store. These are arranged with lots of little compartments which are perfect for separating all your threads, pins and needles, tape-measure and scissors.

Thread comes in all sorts of thicknesses. For the projects in this book, I have almost always used a general polyester thread. The really cheap brands of thread can be a false economy. They invariably break in my sewing machine, which drives me to distraction. If you are sewing with 100% cotton fabric, you can use a pure cotton thread. Likewise if you are sewing a synthetic, man-made material, you should use a synthetic thread.

Pins and needles These always seem to vanish into thin air, and you're bound to break your last needle just at the start of your work! So always have plenty of needles, both for hand sewing and the machine, in your workbox. You will need thicker needles for heavyweight fabrics like velvet or calico, and thinner needles for sewing sheers and muslin. Occasionally I recommend that you use a curved needle for stitching the trimming, for example, onto the pelmet. You will find this is very useful and it will certainly be kinder to your

fingertips. It is also useful to have several embroidery type needles with wide eyes and with both blunt and pointed ends. I have sewn the decorative ribbonwork through the pretty baby quilt using a blunt needle.

I think dressmaker's pins are preferable for the projects in this book as the thick ones will break threads and leave holes in your fabric. Coloured headed pins are also a boon for clumsy sewers as they are so much easier to find on the carpet.

Tapes Always keep a cloth tape and an expandable metal tape in your basic tool kit. You will use either one for almost every task.

Seam ripper This is an extraordinarily cheap piece of equipment but I would be lost without mine. It's very sharp and unforgiving if you are in a hurry, but used with care it will unpick all your mistakes quickly and easily.

Scissors A pair of good, sharp dressmaking scissors is really the only other essential piece of equipment you'll need to make the projects in SIMPLY FABRIC. But try to save them just for sewing. If you don't, you'll find that blunt scissors, used for everything from cutting paper to pruning the roses, will reduce your favourite piece of lace or fabric to ribbons! I also have a pair of pinking scissors which I find a useful addition to my tool kit although they are not essential. On certain fabrics they can be used along the seam allowances to stop the cloth from fraying.

MATERIALS

I could happily spend hours wandering through showrooms that display reams and reams of glorious fabrics. Funnily enough I always seem to be drawn to the beaded silks and shimmering organza. And the only thing that prevents me from using the more luxuriant velvets and damasks is their price. Otherwise I have no limitations. Many of the fabrics I have used in SIMPLY FABRIC are from well-known suppliers. However if you have difficulty obtaining them, you can use any fabric that catches your eye, provided it is of a similar weight and texture. I almost never adhere to the hard and fast rules applied to using fabric, as I feel that the most important thing is to get excited about a brightly patterned or unusual piece of cloth. I'll often buy something that is bold enough to catch my eye, and only later find a use for it. Sometimes I'll rediscover a fabric at the very bottom of my storage trunk, which I bought ages ago and never used, and I will instantly be inspired and know just what to make with it. So, it can be worthwhile buying that old piece of lace from an antique shop, or the odd remnant from a designer sale, in the hope that one day it will inspire you.

In the instances where I have painted the fabric myself, say for the primulas, any inexpensive, plain fabric can be used – I used up all my scraps of left-over curtain lining. I have practised these projects using other material, including silk, but as the glue hardens the fibres in the fabric, the effect is just the same. Your choice of fabric often depends on where in the house you ulti-mately want it displayed. For a sunny window requiring an element of screening and just a little privacy, it would be a mistake to hang a dark, heavy fabric, unless you simply hate the sunlight, of course! It would be much better to use a sheer fabric that didn't block available light, but which at the same time, gave you privacy during the day-time. You could then pull heavier curtains over these at night. Sheer fabrics also have the added appeal of being remarkably inexpensive, so it is possible to gather up luxurious folds and swags of muslin for a sumptuous window dressing, saving the more costly fabric for blinds that only require enough fabric to cover the window.

Which fabric? Flowers, stripes, checks? There are so many fabrics that it's sometimes impossible to decide on any one in particular. I would say that the most important factor is yourself. At the end of the day, you will have to live with your choice, and my preference for orange polka dot curtains will almost certainly not be everyone's cup of tea!

However, there are a few things that you can do to make these choices a little easier. If you are buying a large amount of fabric, for say a window treatment, take sample colours of other upholstery or fabrics used in the room on your shopping spree. If you can also take a sample of the carpet and a dab of paint on a card, these will also help you to make the right choice. It's a good idea to look at fabric in natural daylight so take the roll of material to the shop window to view the texture and colour. It's surprising how much the colours do change. Some fabric suppliers will allow you to take large samples of fabric home, which is an excellent idea. The cost may then be refunded when you return the fabric. Depending on what the fabric is to be used for, give it the scrunch test. Simply crumple a corner of the fabric in one hand and see to what extent the fabric remains creased.

Some fabrics are more crease-resistant than others, so avoid using something like linen for cushions or bedcovers which would soon look crumpled and messy. It is important to know what I mean by the weight of the fabric, so you can change needles and machine tension accordingly. Lightweight fabrics include muslin and voile and all types of sheers. Medium-weight fabrics include most of the cotton prints and chintzes; the heavy-weights include canvas and calico.

Lining Some of the lining fabrics I use include wadding, which is a padded, usually synthetic, interlining. I use this extensively to soften an otherwise hard surface such as the pelmet, headboard or the lid of the toy box.

Buckram is a heavy stiffened cloth that is usually made from cotton. This is particularly useful for lining tiebacks and curtain pelmets.

Curtain lining is usually cotton sateen, which comes in the instantly recognisable buff or ivory colour. However I often think it is better to find a coloured lining for curtains such as the many glazed chintzes, which look good and come in hundreds of colourways. Or why not think about lining a patterned fabric with another patterned fabric? It's daring and a little expensive, but can look very dramatic and luxurious.

Sewing tips The easiest way of joining two pieces of fabric is to sew a flat seam using the straight stitch on the sewing machine. Sew a straight row of stitches, leaving an even seam allowance down the right hand side. Press this seam open using a steam iron. If you are working on a curved edge or point, you will find it easier if you carefully clip the fabric allowance several times with a pair of sharp scissors, almost up

to the line of stitching, before turning the fabric right side out. This will make the finished work lie flatter and look neater.

Fabric adhesive As a general rule I always use PVA craft glue to stick fabric. It usually dries clear, which is good news for non-sewers, and it is ideal for gluing fabric to wood or MDF (medium density fibreboard) as I have on the folding screen. Rubber solution glue is another excellent adhesive, although it is not as strong.

Trimmings I love trimmings as they can embellish the plainest of fabrics and transform them beautifully. Throughout the book I have used various trimmings to add to the overall design. I have made these myself and given you the instructions, but these are a real labour of love and you may prefer to buy ready-made trimmings.

There is really no excuse for being lost for fabric ideas. After all there are hundreds of design books and magazines on the market. My inspiration comes from leafing through them all! I also love visiting showrooms and browsing around department stores, all of which have a wealth of good ideas that can be copied at home! We all get our inspiration from somewhere – I hope that you get yours from S I M P L Y FABRIC

CHAPTER 1

NO SEW

This chapter is full of wonderful fabric ideas, and you don't even have to pick up a needle and thread to achieve the brilliant results! We often think of fabric makes as being tricky and time-consuming, but they're not all about working with reams of expensive fabrics that are too precious even to cut. You'll soon discover how to create stunning projects from fabrics you thought were good for nothing but the dustbin. It only takes half a metre of inexpensive cotton and a little glue to transform your bare kitchen dresser with my lacy shelf edging. Or you could indulge your romantic fantasies by creating the perfect fairytale bedroom with my sheer fabric panelling. Quick and easy ideas are what this chapter is all about, designed for people who just want to spend their time enjoying the end results.

Hatboxes

The three patterned hatboxes are actually some old wooden ones I found in a junk shop, but you can make your own from cardboard like my decoupage hatbox.

MATERIALS

cardboard

masking tape

wadding

PVA

canvas fabric

scissors

piping cord

flowered fabric

Hints

Use cardboard boxes from supermarkets as the basis for your hatbox. To ensure that the curved areas are as smooth as possible, roll up the cardboard to create the shape before you assemble each section.

1 Decide on the diameter of the box then cut a round base to fit. Cut out sides about 30cm/12in deep and long enough to fit the base. Cut another slightly larger circle of cardboard for the lid and 5cm/2in deep strips for the side of the lid. Use masking tape to join the pieces together.

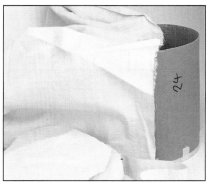

2 Cut a piece of wadding to the same size as the lid and stick down using a little PVA glue. Cut canvas fabric to cover all the box, excluding the base, allow 1.25cm/½in all around for turnings. Remember to allow extra for the sides of the lid.

3 Use PVA to stick the canvas to the box making sure you smooth out the fabric carefully to eliminate any creases. Use sharp scissors to clip the edge on the lid piece to enable it to fold over and fit around the sides.

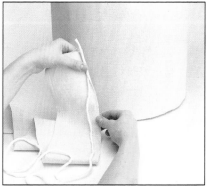

4 Cut a 10cm/4in wide strip to fit the circumference of the lid. Put a little glue along the top edge and use your fingers to roll a length of piping cord inside this edge making sure the raw edges are concealed.

5 Glue the rolled edge of the strip to the side of the lid close to the top edge. Glue down the remainder of the strip and fold, then glue the turnings to the inside of the lid. Leave to dry.

6 For decoration I have cut out flowers from a piece of patterned fabric and used PVA to stick individually around the bottom edge of the hatbox creating a pretty decoupage effect.

Blanket Box

You can often find old blanket boxes or ottomans in second-hand shops, so watch out for the bargains.

MATERIALS

fabric

wadding

pattern paper

PVA

scissors

staple gun/tacks

tape measure

1 Measure lid, sides and plinth of box. Add 1.25cm/½in all around to the sides and plinth for turnings and an extra 4cm/1¾in around the lid. Transfer the dimensions to pattern paper. Cut out the templates and then the fabric pieces.

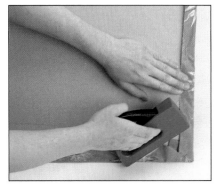

2 Cut a piece of wadding to the exact size of the lid to make a softly padded top. Use a little PVA to glue this in place on the lid. Position the fabric over the wadding on the lid and fit loosely to the inside of the lid with a staple gun or small tacks.

3 Cover the front of the box with PVA glue. Position the fabric, smoothing out from the centre. Turn the excess fabric to the inside of the box and around to the two sides. Glue in place. Repeat for the back. When covering the sides, fold in turnings and glue down for a neat edge. Paint any decorative details in a contrasting colour.

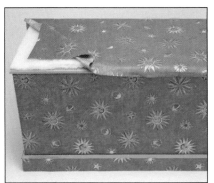

4 Use a staple gun or tacks to firmly secure the fabric to the lid, keeping the fabric tight throughout. At each corner fold the fabric carefully to achieve a mitred finish. Cut a piece of fabric to line the lid. Fold in turnings and glue in place.

Pleated Lampshade

This simple shade is made from a lightweight fabric glued onto thin pattern paper. You can use a plain or patterned fabric to match or contrast with your furnishings.

MATERIALS

lampshade frame

fabric

scissors

PVA

pattern paper

pencil

ruler

hole punch

ribbon

all-purpose adhesive

Hints

A touch of glue, at regular intervals, along the inside bottom edge of the shade will hold it in place on the frame.

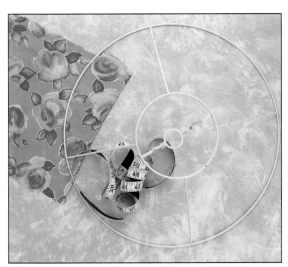

1 Measure around the circumference of the lowest ring and multiply this by one and a half to give you the length of your fabric. Measure a vertical strut to give you the width of your lampshade. This is one project where you do not need to allow for turnings.

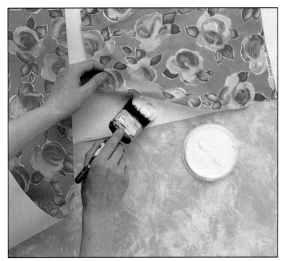

2 Thin PVA with an equal quantity of water and use to glue the fabric to the paper. Leave to dry, then cut out the rectangular shape from the calculated measurements. Use a ruler and pencil to draw vertical lines on the paper at 1.25cm/½in intervals. Score every other line with the edge of a ruler.

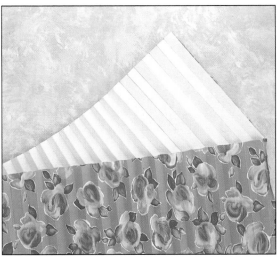

3 Fold the shade in concertina pleats using the scored lines to help you line up the pleats and ensure that they are even. The two ends should neatly overlap each other. Use a strong all-purpose adhesive on the overlap to conceal the join.

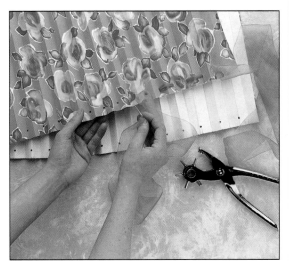

4 Use a punch to make holes in the top of the shade at regular intervals, about 1.25cm/½in from the top edge. Thread fine ribbon through the holes. Position the shade over the frame to determine the right size of the shade. Pull the ribbon tight and secure with a neat bow.

Leafy Picture Frame

This is almost a work of art in itself so if you want you could hang the frame with no picture at all!

MATERIALS

PVA

plain fabric

tracing paper

pencil

pattern paper

scissors

gold spray paint

Hints

Depending on the colour of your fabric you can use different spray paints to create different effects to suit your colour scheme.

1 Thin PVA glue with an equal quantity of water and use it to paint all over the fabric. Leave it to dry and so stiffen the cloth. Almost any kind of fabric would work as the glue makes even the finest fibres hard.

2 Trace our leaf shapes and make a paper template from which you can cut lots of fabric leaves. Turn the leaves around to save wasting the stiffened fabric.

3 Fold each leaf shape in half lengthways and hold between your finger and thumb. Use gold spray paint to colour the tips of the leaves allowing the colour of the fabric to show through. This gives an interesting shaded effect.

4 Use PVA or any strong adhesive to stick each leaf onto the frame. Follow a regular pattern as I have done to build up the frame, adjusting the leaves slightly to go around the four corners.

Lacy Shelf Edging

As open shelving is so popular in country-style kitchens, this is an attractive way to finish the cottage look. Use glue or decoratively headed tacks to position the lacy edging on to your shelf or dresser

MATERIALS

PVA

household paint brush

plain fabric

plastic

pattern paper

scissors

tea cup/ruler

pencil

pinking scissors

fabric punch

1 Thin half a cup of PVA glue with an equal quantity of water. Use a household brush to paint this over one side of some plain fabric to stiffen. You will find it easier to put a layer of plastic underneath the fabric. When the fabric is dry this will simply peel off the stiffened cloth.

2 Cut a paper template using the curved rim of a teacup to form a scalloped edge or a ruler to make an easy zigzag shape. Draw your pattern onto the fabric, repeating it until your trim is long enough to fit your shelf. Cut out the edges using pinking scissors for a pretty lacy quality.

3 Draw a simple design on one of the shapes. Fold the shapes over one another, in a concertina fashion until you have a thickness of four or five. Use the fabric hole punch to form the pattern. Vary the size of the holes for a more interesting effect.

Fabric Primulas

These pots of flowers look fabulous, particularly if there are two or three grouped together. Show them off in a small bathroom where it is difficult to grow fresh plants.

MATERIALS

coloured acrylic paints

PVA

household paint brush

white cotton fabric

tracing paper

pencil

pattern paper

scissors

artist's paint brush

florist's wire

florist's tape

skewer

small flowerpot

florist's foam

sphagnum moss

Preparation

Select the coloured acrylic paints for the flowers. Mix half a cup of PVA glue with a 5cm/2in squeeze of the appropriate colour. Use a household brush to paint each colour onto a piece of white cotton fabric. Hang the fabric up to dry, preferably outside on a clothesline.

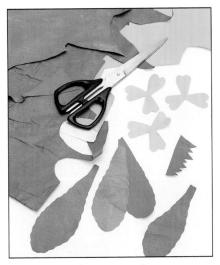

1 Copy our shapes to make paper templates of the petals, leaf and calyx. Cut two petal shapes for each flower from the yellow fabric and ten leaf shapes for each pot from the green fabric and one calyx for each flower from the green fabric.

2 Shade the centre of each petal shape with green acrylic paint. Different coloured flowers should have different shading. Paint faint white lines on the leaf shapes to represent the veins in the leaves.

3 Prepare the flower stems by wrapping a calyx around the top of the florist's wire, hold in place by binding with florist's tape. Keep the tape taut all the time overlapping it slightly as you twist it down the stem.

4 Pierce the centre of each petal with a skewer. Spread PVA on to the calyx and push two petals on the wire to surround this. Bind florist's wire to the leaf shapes with tape. Put flowers and leaves into a pot with dry florist's foam and cover with sphagnum moss.

Decorative Fans

These fans can be used to decorate the edges of a shelf, but I think they would look very striking fixed under the cornice for an attractive ceiling decoration.

MATERIALS

PVA

pattern paper

fabric

scissors

needle

thread

staple gun/tacks

all-purpose adhesive

Hints

Use a fabric with an obvious check or stripe as you can follow the pattern when you pleat the fan into shape.

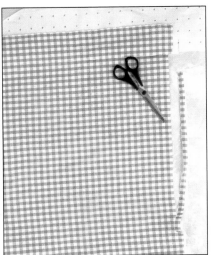

1 Thin PVA glue with an equal quantity of water and use to stick the fabric to a piece of paper. I have used pattern paper to back my fabric as this is easier to fold when dry. Cut 24 x 7.5cm/9½ x 3in rectangles from the stiffened fabric.

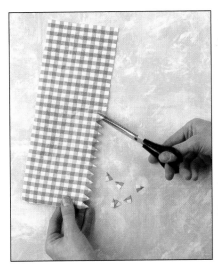

2 Snip along one long edge to form tiny zigzags. If you choose a gingham fabric as I have done, simply use the squares as a guide to cut an evenly spaced hem. Otherwise cut the zigzags in proportion to the design but no more than 1.25cm/½in deep.

3 Again use the pattern in the fabric to determine the size of the folds and pleat the length of the rectangle. Press each fold quite hard with your thumbnail to hold the creases in place.

4 Open the shape and pinch the two inside edges together to form the fan shape. Stitch these edges together by hand to hold the fan open. Use a strong all-purpose adhesive to attach the fans in position to the underside of a shelf or cornice.

Plaited Picture Frame

I love the unique texture of this plaited frame which actually makes something very special out of the quite ordinary.

MATERIALS

felt – 2 or 3 colours

PVA

picture frame

small brush

scissors

Hints

You could use fabric or even string plaits to make another picture frame. Don't forget that any frame, including mirrors, can be transformed in this way.

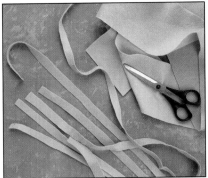

1 Cut the felt into long thin strips. For the larger picture frame on the front cover I used felt strips that measured approx 2.5cm/1in. For a smaller frame cut narrower strips that are about 1.25cm/½in wide. .

2 Plait three strips together in the usual way, until you have ten lengths. You may find it easier to have someone hold the ends while you plait. Cut four pieces of felt to cover the frame, front and back. Remember to mitre the corners. Stick to the frame.

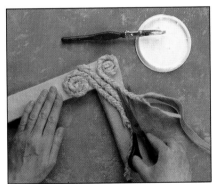

3 I have coiled the plaits around in some areas and formed wavy patterns in others. Whatever design you choose, do make sure you use plenty of the PVA glue underneath each plait. Don't worry if there are small splatters on the surface as this glue will dry to a clear finish.

Folding Screen

Once popular at the turn of the century, screens are now enjoying a decorative comeback.

MATERIALS

fabric

cotton interfacing

PVA

scissors

pattern paper

Hints

Clip the hem allowance to make it easier to fold the fabric evenly over the curved areas. Use undiluted PVA at the edges for a stronger bonding.

1 Separate the panels of the screen remembering the position of the hinges. Use each panel as a template for the fabric panels. Cut fabric for each screen panel, front and back. Add at least a 2.5cm/1in turning allowance around all the sides.

2 Unless your screen is white, any other colour may show through. If so back the fabric with a layer of cotton interfacing. Cut out the iron-on interfacing using the cut fabric pieces as templates. Use a hot iron to fuse the interfacing onto the fabric.

3 Thin the PVA glue with an equal quantity of water and apply to each panel. Stick the fabric in place on each panel, pressing from the centre outwards in all directions, to avoid any air bubbles. Turn over and glue down edges. Back the screen with complementary or contrasting fabric.

Muslin Drapes

These are created without any stitching using three layers of muslin which are tacked or stapled to the top of the wooden architrave that surrounds the window.

MATERIALS

tape measure

muslin

scissors

staple gun/tacks

Hints

If you don't like the idea of white muslin, why not dye it first to a shade that will complement your colour scheme?

1 Measure the window and cut enough muslin to cover it completely. Tack or staple the required number of drops to cover the window, overlapping if necessary but without any fullness or gathering. Use double thickness for privacy.

2 Cut two lengths of muslin three times the depth of the window. Using double thickness of muslin, tack or staple one narrow edge straight onto the window frame on the right. Repeat with the other narrow edge on the left hand side. The muslin will now fall in a huge loop.

3 Gather the loop together in the middle and tie into a huge, soft knot to hang in the centre of the window. This can be tricky but easily achieved with an assistant. Spread the bottom edge to drape evenly along the floor. Cut a length of muslin to the width of the window. Staple or tack along the top of the architrave to disguise the other layers. Gather intermittently from the bottom edge to form loops and tack in place on the architrave. This gives a softly swagged finish.

Fabric Panelling

This unusual decorative treatment using sheer fabrics is perfect for creating the ultimately romantic bedroom.

MATERIALS

sheer fabric

scissors

staple gun

ribbon

lace ribbon

adhesive

Hints

To clean the fabric panels, simply untie the centre ribbon and shake the dust out of the folds and re-tie.

1 The number of panels needed depends on the size of your wall. Each panel measures 40cm/16in wide. To determine the depth, measure the distance between the dado and the skirting, adding 5cm/2in for turnings. Staple or tack the first panel under the dado folding in turning.

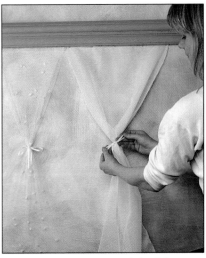

2 Calculate the centre of the fabric panel with the centre of the dado and tie a narrow piece of ribbon around this point. Tie the ribbon securely using a double knot and cut off the ends to make a pretty bow.

3 Gently pull the fabric to the top of the skirting board, fold in turnings and staple or tack along this edge. You will need to ease the fullness as you proceed to make a neat panel. Trim away any excess fabric.

4 Use a flat lace ribbon and a strong adhesive to cover the staples or tacks. This can be as wide or narrow as you want. I have used a narrow ribbon but a wider satin ribbon would look equally effective.

CHAPTER 2

EASY SEW

This chapter is all about getting maximum effect from minimum skills. So don't be daunted by any of the projects; they are all far easier to create than you may think. Some of the most stunning designs are often those that are the simplest to make. After all, you probably know about the basics of sewing, even if it's just mending a hole in your best jeans! Most of these projects don't involve anything more complicated than a simple running stitch, sewn either by hand or on a machine. I'm sure that by the time you've tackled a couple of ideas from this chapter you'll be encouraged to attempt some of the more ambitious projects later in the book.

Firescreen

I found this screen in a second-hand shop. The centre section was so shabby it had to be replaced. I have painted the frame using two colours of emulsion paint and then aged the surface by rubbing with sandpaper.

MATERIALS

firescreen

2 colours emulsion paint

paint brush

sandpaper

fabric

scissors

thread

curtain wire, rings

hooks

Hints

The two lines of stitches that form the casing should be as close together as you can make them. Use the width of the ring on the curtain wire plus a little extra as a guide.

1 Cut two equal pieces of fabric to the size required plus 2.5cm/1in all round for seam allowance. Use different fabrics on each side making sure they can be washed together. Right sides facing sew the two fabrics together, leaving a gap for turning. Turn right sides out and sew up the opening.

2 Decide at what point you want to position the curtain wire. Sew one row of stitching directly above this point and another directly below it to make a narrow channel that is slightly wider than the ring on the curtain wire. Repeat for the bottom edge.

3 Cut two pieces of curtain wire to the required length. Screw the rings into each end. Use sharp scissors to cut the few stitches in the seam between the casing lines, and thread the curtain wire through. Screw four curtain hooks into the wooden framework of the screen in positions to correspond with the rings on the fabric cover. Pass the four rings over these to secure the fabric panel to the screen.

Place Mats

I like this quick and simple technique and have made several sets of mats for different occasions.

MATERIALS

fabric

matching thread

scissors

dinner plate

Hints

You don't need to make all your mats in the same colour. Sometimes they can look more interesting in mismatched colours.

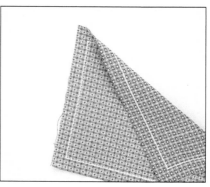

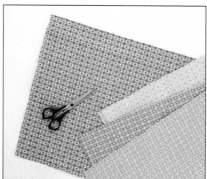

1 Choose a loosely woven fabric that will fray easily. Use a large dinner plate to determine the basic size of your mat then cut a rectangular shape, in the fabric, at least 10cm/4in larger all round the plate, so that it will fit easily within the mat.

2 Using a zigzag stitch on your sewing machine sew a row of stitches 2.5cm/1in in from all four cut edges. This forms a tight edge and will allow you to fray the fabric up to this point.

Napkin Rings

I have designed these to match the place mats, but you could make them in contrasting colours and fabrics.

MATERIALS

fabric

matching thread

needle

scissors

napkin

Hints

You could use a contrasting thread in your sewing machine to form a coloured stitching seam for an informal look.

1 Roll up a napkin to give you the diameter of the ring. Add 4cm/ 1¾in seam allowance. The width of the ring should be 8cm/3¼in. To this measurement add 2.5cm/1in for the frayed edge to match the place mats. Using a zigzag stitch on your machine sew a row of stitches 2.5cm/1in in from the long, cut edges.

2 Use a straight stitch to sew the two narrow edges together to form a circle. Cut one edge of the seam close to its stitching line. Fold the second edge over by 1.25cm/½in. Fold this over the seam to enclose it. Hand stitch in place using a needle and thread.

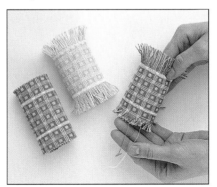

3 From the edge of the ring carefully tease out the threads until you have reached the row of stitches. You won't be able to pull any more threads beyond this point. Continue until the top and bottom edges of the napkin ring are frayed evenly.

3 Use your fingers to tease out the horizontal threads around the edge of the mat. Remove all these threads until you have reached the stitched seam on all four sides. You will now have a beautifully soft, frayed border.

Blind Trim

This is a wonderful way of jazzing up your existing plain blind for a small cost. Simply sew the strip down the centre of the blind.

MATERIALS

fabric – main border

fabric – coloured strip

scissors

thread

iron-on decoration

Hints

An open weave fabric is ideal for this treatment as it will fringe very easily.

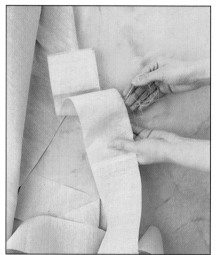

1 Cut strips of fabric, 7.5cm/3in wide, to form the border. Sew strips together, if necessary, to achieve the same length as your blind. Allow 4cm/1¾in for the turnings at the top and bottom. On each of the long edges tease out the threads to give a softly frayed border.

2 Cut the coloured fabric for the centre strips. These should measure 7.5cm/3in including turnings. Join strips if necessary. Fold the fabric over and sew a seam down the length 1.25cm/½in from the edge. Trim the seam and press. Turn right side out.

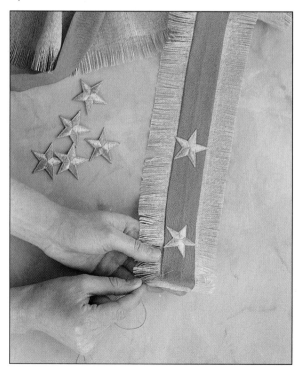

3 Sew the fabric piece to the centre of the border down both sides, using a straight stitch. Apply your decoration. I have used iron-on gold stars but you could use anything from tiny shells to pretty buttons or beads. Sew to centre of the blind.

Fabric-covered Pelmet

Using MDF (medium-density fibreboard) you can create any shape of pelmet you wish, from a simple scalloped shape like mine to something curved and fanciful.

MATERIALS

pattern paper

scissors

wadding

fabric – patterned

fabric – white

PVA

curved needle

thread

wooden beads

emulsion paint

paint brush

all-purpose adhesive

Hints

You can use ribbon for your gathered trim. Any ribbon would work provided it is wide enough for the two lines of gathering stitches.

1 Make a paper template of the pelmet. Cut wadding to the same size. Stick the wadding to the pelmet with PVA glue. When the fabric is stretched over the edges it will flatten the wadding and this will soften the hard edges of the pelmet.

2 Cut the fabric to fit the pelmet allowing 2.5cm/1in extra all round for turnings. Cut notches in the edges of the fabric so it will fit around the curves easily. Glue excess to the back. Cover the back of the pelmet folding in the turnings and gluing down for a neat finish.

3 Cut 5cm/2 in wide strips from the white fabric, join if necessary until it measures twice the length of the pelmet. Sew two rows of long stitches, 1.25cm/½in from each edge. Pull up to gather the trim.

4 Use a curved needle to sew the gathered trim to the pelmet. Sew a row of stitches down the middle of the trimming. Catch plenty of the wadding with each stitch as the trimming needs to lie flat against the pelmet.

5 As an optional extra you can paint wooden beads with emulsion paint and secure with adhesive, at intervals along the gathered trim.

Bath Mat

Stepping out of a bath onto a
really soft mat is, I think, the
finishing touch to bathtime
luxury.

MATERIALS

towelling

thread

needle

scissors

Hints

When you plait towelling, which is quite
bulky, you will get a better result if you
work quite loosely – your mat will also
come together much quicker.

Preparation

Cut the towelling into nine strips
approximately 15cm/6in wide by
90cm/36in long. Sew the lengths
together so that you will have three
long lengths. It is a good idea to shake
the cut towelling at this stage to
remove any loose threads.

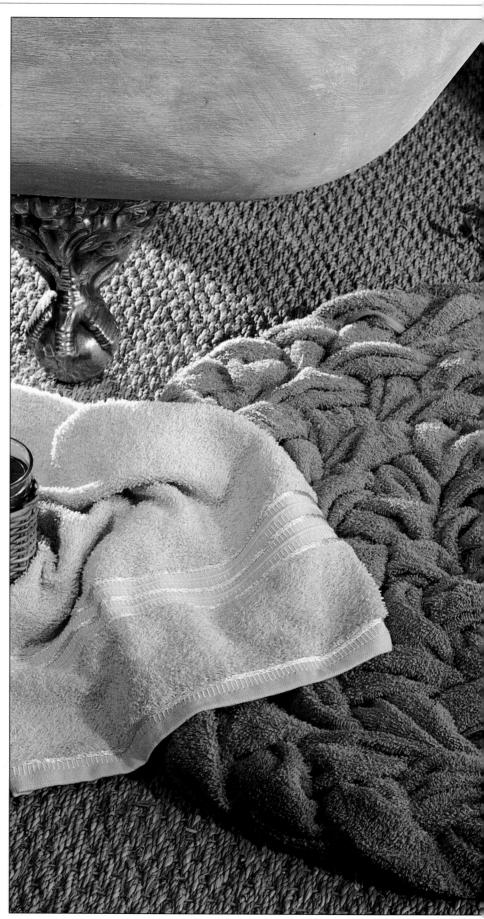

1 To prevent the side edges from fraying oversew using the long zigzag stitch on your sewing machine. This may seem rather time consuming but it is worth it because otherwise you will find that your mat will start to disintegrate the first time it goes into the washing machine.

2 Place the three strips on top of each other and sew together along the top edge. Plait the strips in the usual manner, bunching the towelling and turning the edges underneath as you do so. It may be easier to work on the floor at this stage. Secure final edge.

3 Carefully turn the plait over. Coil one end around into an oval shape. Using a large needle and strong thread, sew the coiled edges together. The stitches can be quite large as you will not be able to see them. Continue coiling until you have finished. Tuck in the end and sew securely.

Tie-on Chair Cover

This attractive cover is made from three fabric pieces and is gloriously simple to put together.

MATERIALS

pattern paper

fabric

scissors

fuseable webbing

pins

thread

needle

ribbons

Hints

If you are in a hurry, all these seams can be turned up using fuseable webbing. You can secure the ribbons in this way too, although the cover may not survive the washing machine.

1 Cut three main pieces from pattern paper, two sides and one long piece to run from the front hem all the way over the seat and backrest to the back hem. Using the paper pattern cut out the fabric and add 1.25cm/½in all round for a seam allowance.

2 Fold over the seam allowance along the sides that are not sewn and use an iron to press flat. Sew or use fuseable webbing on the seams. Pin, then sew the two side pieces to the seat and front edges. These are the only seams to be sewn together.

3 Turn the cover to the right side and stitch pairs of ribbons opposite each other, at regular intervals all the way down the open seams. Tie the ribbons into neat bows and trim the ends to finish.

Animal Cushions

These brightly painted cushions can also be used as toys. An older child can paint the animals using coloured fabric paint. Remember to seal the paint by ironing.

MATERIALS

pattern paper

pins

calico fabric

scissors

filling

needle

thread

fabric paint

artist's paint brush

Hints

Enlarge your child's animal drawings on a photocopier and use as patterns for the cushions.

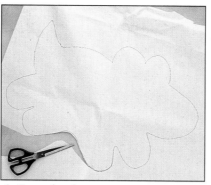

1 Draw the shape onto pattern paper. Your sewing line will be 1.25cm/½in inside this line. Remember the simplest shapes often look best. Pin the paper template onto a piece of double thickness, calico fabric. Cut carefully around the outline.

2 Sew all around the shape keeping your seam allowances even. Leave an opening of 10cm/4in for turning the shape through. Snip the curves almost up to the sewing line to ease the corners.

3 Turn the animal shape to the right side and stuff firmly using a suitable filling. Take care when choosing your filling, you will find several safe hygienic fillings on the market that are suitable for small children. Sew the opening together.

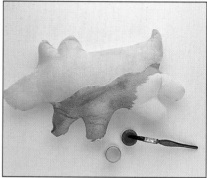

4 Using brightly coloured fabric paints, shade in the animal details onto the calico shape. These can be as realistic or as fanciful as you wish.

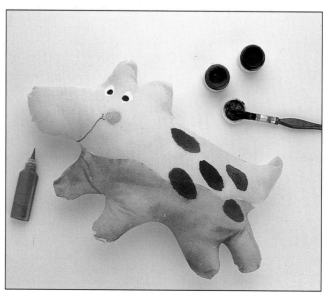

5 Use your imagination and over emphasise all the facial markings for a dramatic look. I used blobs of paint on my fingertips to create the eyes; a tube of fabric paint with a nozzle is ideal for painting the animal's smile.

Tie-on Blind

Decide whether the blind is
to hang inside or outside the
frame, then use these
measurements to calculate
the size of the blind.

MATERIALS

fabric – main

fabric – border

scissors

touch-and-close tape

pins

ribbons

needle

thread

Hints

This blind hangs from a narrow wooden
batten fixed to the top of the window.
It is held in place by a touch-and-close
fastening tape. The soft half is sewn to
the blind and the remaining half of the
tape is glued to the front of the batten.

1 Cut the main fabric to fit the window. Add 2.5cm/1in to the length for turnings. Cut three 6cm/2¼in strips for the borders, to cover the length and width of the blind, add 5cm/2in for turnings. Fold in a narrow hem on the longest edges of the borders. Right sides facing pin then sew to the sides and base of the blind. Fold over the raw edge, sew in place on the back of the blind.

2 Cut off the excess fabric on the border at the base of the blind. Fold in the two ends of the borders on each side and position over the border at the base of the blind to sandwich it and create a neatly mitred corner. Hand sew in place.

3 Fold over the top edge by 2.5cm/1in. Over this sew the soft half of a touch-and-close fastener. From the contrasting fabric cut four ribbons 10 x 45cm/4 x 18in. Fold in half and sew together down longest edge. Turn right sides out and neaten ends. Sew to the top of the blind at the same point front and back. Gather up the blind to the height required and tie the ribbons.

Pointed Tablecloth

This technique may be used to create any number of decorative hems. Our larger cover has a hem that is cut with the pattern on the fabric and the smaller tablecloth has a pointed edge.

Hints

The fabric interfacing used to back the tablecloth has a woven cotton backing which is stronger than most soft interfacings.

1 Measure the circumference of your table, divide this measurement into equal sections; ten points should be sufficient to surround a small table like ours. Add 5cm/2in to the width for a generous seam allowance. Cut a paper template from these measurements then use to cut the fabric and interfacing.

2 Along the top edge of the fabric, fold over 1.25cm/½in. Press down with an iron. Iron the interfacing to the wrong side of fabric to almost cover this fold, ensure that there is a tiny strip of the fabric showing at the top.

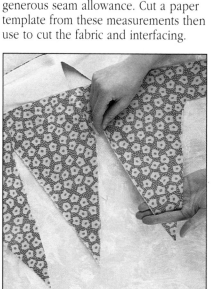

3 Use the smallest zigzag stitch on your machine to finish the raw edges. I have used a contrasting thread so you can see how close the stitching is but, of course you should choose a thread that matches the colour of your fabric.

4 Cut a piece of fabric to the diameter of the table adding 2.5cm/1in seam allowance. Use a zigzag stitch to finish the raw edges. Place fabric on the table and pin the pointed edge around the circumference. Remove to hand stitch the two pieces together.

Three Sheer Window Dressing

This is an ideal way to cover a small window. It is not designed to be drawn open but the sheer fabric filters the light, while the beads and gold thread sparkle for a very pretty effect.

MATERIALS

organza fabric

sheer fabric

shiny fabric

gold thread

card

sewing thread

needle

glass beads

pearls

dowelling

Hints

Secure the dowelling rods to the wooden architrave around the window, use light-weight brass fittings especially designed for curtain rods.

1 Cut all three fabrics to the size of your window. Add 2.5cm/1in to each side for turnings and 7.5cm/3in to the top edge to make a channel heading for dowelling. Allow 2.5cm/1in along the bottom edge for a hem.

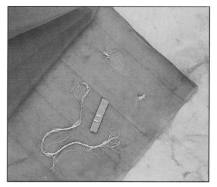

2 To decorate the organza, wind gold thread approximately ten times around a small piece of card. Slip it off the card and use more gold thread to tie a knot around the middle to make a tiny bow. Repeat this until you have enough bows to stitch at random all over the fabric.

3 The second silky fabric has a beaded decorative edge. For this I restrung small beads from old necklaces. Use pearls and glass beads for a pretty effect. Make the bead lengths about 5cm/2in long. Sew these to the bottom edge of the fabric after hemming.

4 I have sewn small individual pearls onto the third piece of fabric but you could use any type of beads. Glass ones look good but be careful not to use very large ones as they will pull the fabric and spoil the final display.

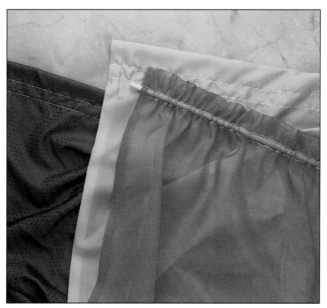

5 Sew all the side seams and bottom hems. The fabric with the beaded hem should be the same length (including beads) as the other two. Make any adjustments at the top edge. Fold over seam allowance at top edge and sew down. 2.5cm/1in above this sew another row of stitches to form the casing for the dowelling.

Ribbonwork Cushion

The simplicity of this cushion cover reminds me of the weaving I used to do as a child and yet the design looks quite sophisticated.

MATERIALS

4 patterned or plain ribbons

cotton fabric

pins

scissors

thread

cushion pad

Hints

Make your cushion cover a little smaller than the pad to give your cushion a plump look.

Preparation

Choose patterned and plain ribbons of varying widths for the best results. Contrasting colours are strong visually so I have used orange and red with green, but lilac and yellow would be equally vibrant. Cut the cotton fabric to the same size as your pad.

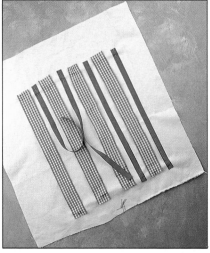

1 Pin the first two ribbons onto the cotton backing keeping an even spacing between the edges. If you pin at right angles to the ribbon edge you will be able to sew over the pins without having to remove them first.

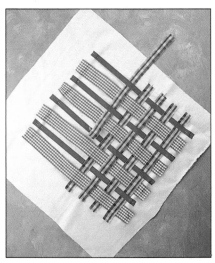

2 Take the third ribbon and weave this between the first two ribbons. Weave over the first ribbon then under the second ribbon. The next weave will then start by going under the first then over the second. Continue until the third colour is finished.

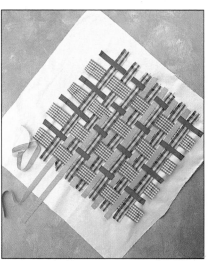

3 Follow the last procedure to weave the fourth ribbon, until all the gaps have been filled with ribbons. Remember to pin each ribbon at right angles every time you complete a row.

4 Cut two more pieces of cotton to the pad size. Place one square over the worked ribbons. Fold back a third then pin and sew to the cushion top around the three sides. Repeat with the second square to form an envelope opening. Turn through to the right side.

Tassels

Use these tassels to tie back
your curtains or to decorate
cushions or pelmets.

MATERIALS

wooden beads – 3 sizes

glue

card

scissors

crochet thread

needle

Hints

For decorative effects you can use
different shades of crochet thread to
match all the colours in your room.

1 Glue the beads with the smallest at the top. Measure then double the height, add 5cm/2in. Cut a piece of card to this length and 10cm/4in wide. Wind crochet thread round the card 50 times. Remove from card. Use end thread to wind around threads to form a tassel.

2 Place these tassel threads over the beads and spread out evenly to surround beads. Gather the threads below the bottom bead and again bind with more thread and knot securely.

3 Wind more thread around the two points where the beads join, to create the top of the tassel. Use the same piece of card to create at least twelve more tassels as before. These should all be the same thickness but could be in different colours.

4 Sew these tassels around the bottom edge of the largest bead in the tassel top, spacing them evenly to give a full effect. Attach a triple loop of threads to the top of the beaded tassel, winding more thread around the base of the loop to secure.

Fabric-covered Curtain Pole

This treatment brings a new lease of life to an old curtain pole and will give a completely new look to your window.

MATERIALS

curtain pole

finials

fabric

scissors

thread

needle

PVA

Hints

You can make your own curtain pole from a wide piece of dowelling and two finials, to match or contrast with your curtain fabric.

1 For each finial cut a circle of fabric at least 20cm/8in diameter to wrap generously around the outside of the finial. Place the fabric right side down and then put the finial in the centre of the circle.

2 Pleat the fabric up around the sides of the finial, to draw in the thickness of the fabric. At the point where the finial narrows, take some strong thread and wind it around the fabric to secure the fabric and to form a ball.

Fabric Drapes

These are designed simply as a window dressing and used with the fabric covered pole they create a stunning frame for any window.

MATERIALS

fabric

curtain pole

string

scissors

pins/thread/needle

Hints

Drape the fabric over the pole teasing the fullness out as you require. If you feel that the drapes need a little security you could sew the drapery folds together at several points along the length of the pole.

1 Place the pole at the window and estimate the length of the fabric drape by winding string around the pole in the same way your drape is to hang. Cut two pieces of fabric to this length, allowing 1.25cm/½in all round for seams. Right sides facing pin and sew these pieces together. Leave a small space to turn right side out. Sew opening. Press.

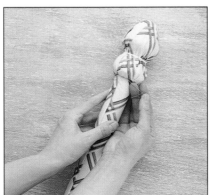

3 Cut fabric to the length of the pole and the width of the circumference of the pole. Add 1.25cm/½in for turnings. Wrap fabric around the pole. Turn under one edge by 1.25cm/½in, overlap the raw edge. Sew along this seam. Glue or stitch the ends.

4 Push the covered finial onto the pole keeping the raw ends tucked inside and stitch the finial to the pole.

CHAPTER 3

QUICK STITCH

There's no way you'll be able to resist the ideas I've designed for this chapter, just let your imagination go wild! Why shouldn't you have tea towels hung at the kitchen window and fabric flowers stuck all over a sunbed if that's what you want? We're still working with the basic items such as curtains and easy tiebacks, but with a little ingenuity you'll find it's not difficult to make them considerably more exciting than all the others. We all need a shower curtain, but why does it have to be boring plastic when you can drape it with luxurious fabric and decorate it with coloured shells? All it takes is a little imagination for a remarkable transformation.

Sunbed

Don't throw out your old sunbed, follow these steps to give it a new and colourful lease of life.

MATERIALS

calico fabric

scissors

black and coloured felt

tracing paper

pencil

pinking scissors

eyelets

needle

thread

Hints

You can use an eyelet kit to fix the fabric flowers to the cover for an unusual look, or stitch the flowers in place by hand through their centres.

1 Cut the fabric to twice the width of your sunbed, add 2.5cm/1in seam allowance. Fold fabric in half and sew together down one long side. Turn right side out. Fold then sew seam allowance on each short edge. Cut 5cm/2in wide supportive calico webbing strips and sew on at the point where the cover is fixed to the frame.

2 Trace our flower shapes and use as a template to cut flower shapes from coloured felt. I have used black felt for the flower shapes and lilac felt cut with pinking scissors, for the round centre.

3 Attach the flower shapes to the cover using eyelets. If you are hand stitching, sew through the centre of each flower. I have used eyelets to attach the top and bottom of the cover to the frame but use the same construction as on your bed.

Ticking Cushions

These cushions are lined with a brightly coloured fabric which will still be visible even when the cushion pad is inside.

MATERIALS

ticking fabric

plain fabric

scissors

pins

needle

thread

cushion pad

Hints

Cover the cushion pad with a different fabric to create a more colourful effect.

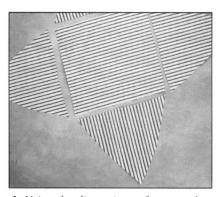

1 Using the dimensions of your cushion pad as a guide, cut out one square and four triangles in both the ticking and lining fabrics. Add 1.25cm/½in for turnings. The four triangles are cut from a square that is 7.5cm/3in wider all round than the basic square. This is then folded across both diagonals and cut along the folds.

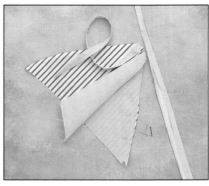

2 Make simple ties for the corners by cutting four 3.75 x 30cm /1½ x 12in strips. Turn the raw edges to the inside, fold in half and sew the seam. Pin each tie between the ticking and the lining of one corner of each triangle and sew in place around two sides, leave the bottom edge open.

3 Turn each triangle to the right side. Pin, then sew one to each of the four sides of the ticking square, keeping raw edges even and the right sides of the ticking together.

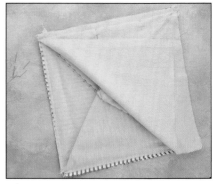

4 Fold all four triangles to the centre of the cushion and pin the square piece of lining over the top, right sides facing and raw edges matching. Sew in place leaving a small gap to turn through.

5 Pull the cushion cover through to the right side and using a needle and thread, hand sew the lining closed. Cover your cushion pad, place in the centre of the ticking cover and join the four ties with a pretty bow.

Tea Towel Blind

Traditional linen tea towels make a very effective blind as their stripes complement the horizontal folds.

Hints

To fix a Roman blind at the window, attach a wooden batten to the frame. Stick the second half of the touch-and-close tape to the front of this. Fix screw eyes underneath the batten in corresponding positions to the rings on the blind.

1 Measure the width and drop required to cover the window. Sew the towels together to this finished measurement. If you need to piece the towels together, remember to oversew the raw edges so the cloth does not fray.

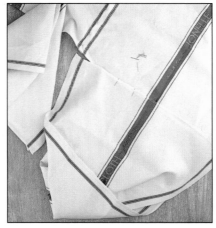

2 Cut 9cm/3½in wide strips of plain fabric to make the casings for the dowelling rods. These should be cut to the same width as the blind plus 2.5cm/1in for the casing end hems. The size of your window will determine how many dowelling rods you will need. The space between each should not be more than 40cm/16in to produce the most attractive folds.

3 Lay the blind flat and leaving a distance of 5cm/2in from the bottom, and 23cm/9in from the top, carefully measure the position of the dowelling casings. Pin then sew the casing to the blind ensuring the seams are perfectly horizontal and that the dowelling will fit in easily.

4 Sew the touch-and-close tape to the top of the blind along both top and bottom edges of the tape. Sew the cord rings at intervals of 50cm/20in to the bottom edge of each casing. Insert rods. Thread the cord through the rings to finish the blind.

Chair/Bed Throw

This throw is warm and comforting with its layer of interlining. Toss it over a bed or an armchair for casual style.

MATERIALS

fabric – top

fabric – lining

interlining

pins

thread

scissors

Hints

Because you are dealing with large pieces of fabric, it may be helpful to tack the layers together, before sewing. Use long stitches which can be removed after the final sewing.

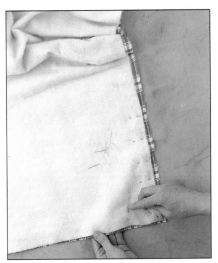

1 Cut top fabric and interlining to fit the top of a single or double bed. Pin and sew the interlining to the wrong side of the fabric. Use a straight stitch and machine approximately 1.25cm/½in from the raw edges.

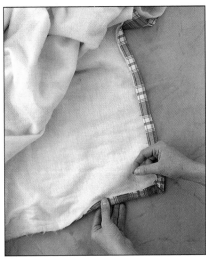

2 Turn the sewn edge over the inter-lining by 2.5cm/1in, then pin and sew this edge to form a narrow hem all the way around the throw. As this is very narrow keep your stitching lines as close to the turned edge as possible.

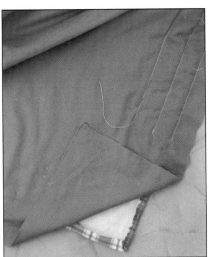

3 Cut the lining 7.5cm/3in larger all around. Pin, then sew to the inter-lined fabric. The excess is to be frayed later. Sew a row of stitches close to the fabric edge. For added detail sew two more rows 5cm/2in apart, towards the centre of the throw.

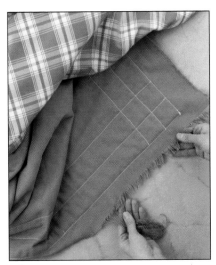

4 Pull out the threads on the excess lining fabric to fray the border. Because the throw is quite large this will take some time but it is well worth the effort. Pull the threads as far as the first stitching line.

Patchwork Curtain

Use an existing plain curtain and sew this patchwork detail onto the bottom edge, or make a simple curtain from cotton sheeting and decorate in the same way.

Hints

I have sewn half a lace tablecloth onto the front of each curtain to make it look even prettier. Sew the tablecloth to the curtain under the heading and let it fall to the point where the patchwork starts.

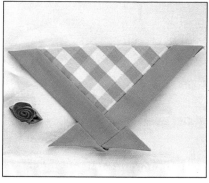

1 The patchwork basket requires three main pattern pieces. Using a ruler draw, then cut the shapes from graph paper. The size of the basket depends on your curtain but try to have at least six baskets on each curtain.

2 Cut one triangle from gingham fabric for the basket shape and two long strips about 2.5cm/1in wide from plain fabric for the basket sides. Two smaller triangles then form the base of the basket. Remember to allow an extra 1.25cm/½in for turnings.

3 Tack the fabric onto the paper template folding over the turnings. Repeat for each of the five basket pieces. Machine stitch along the fold lines to join the pieces together. Pull out the tacking stitches and remove the paper templates.

4 To make the basket handle cut 2.5cm/1in strips across the bias from the gingham fabric. Turn the inner raw edges to the middle. Fold this thin strip in half and sew down the open edge, keeping the machine stitches as close to the edge as possible.

5 Pin all the patchwork baskets to the bottom edge of the curtain. Sew in place with small hand stitches. Sew the ends of the basket handle into the basket leaving the handle free. Position small ready-made ribbon roses at the top of the handle and sew in place with small stitches.

Bedroom Cushions

These are made from sheer voile fabric. They can be patterned or plain but I like to mix two patterns together.

MATERIALS

pattern paper

voile fabric

pencil

string

zip

needle

thread

scissors

pom-pom trim

pins

Hints

To draw a circle fold a square of paper into eighths. Tie a piece of string to a pencil. Cut the string to half the required diameter of the circle. Hold the end of the string with a finger on the point of the folded paper. Pull the string taut and draw a curve from one edge of the paper to the other. Cut out.

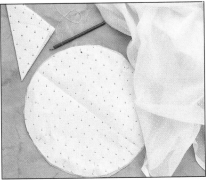

1 Cut a circular template from paper using string and pencil method described in hints or draw around a tray or similar household object. Cut out two circles of fabric.

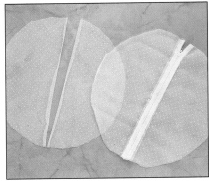

2 Cut one circle in half. Turn under the two central raw edges by 1.25cm/½in. Press. Place the zip face down over the seam, pin and tack the zip in position. Turn the fabric over then sew the zip in place using the zipper foot on the sewing machine. Remove the tacking.

3 Cut strips of voile 7.5cm/3in wide by twice the circumference of the cushion. Join strips if necessary. Stitch two rows of long stitches close to one edge of the strip. Gently pull the threads from one end to gather the ruffles. Sew to zippered circle.

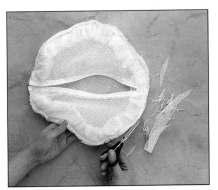

4 Right sides facing lay the second circle of voile over the first. Make sure all the ruffles are flattened to the inside. Pin then sew the top circle in place, matching the raw edges and adjusting the top circle to fit using small tucks if necessary.

5 With the cover turned right side out, hand sew the pom-pom trim to the cushion around the seamed edge, not to the frill. This can be sewn in one long strip or you can cut off the pom-poms and sew them on individually.

Duvet Cover

Duvet covers are very easy to sew and can be decorated with all sorts of trimmings. The scalloped edge gives a neat tailored look to an otherwise plain cover. Use a clean dustbin lid as a guide for the curves.

MATERIALS

3 contrasting fabrics

scissors

dustbin lid

pins

press studs

needle

thread

Hints

If your fabric is not wide enough for the required width of the duvet, join side strips to a central panel for a more professional finish.

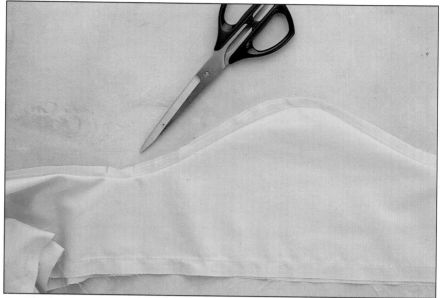

1 Cut two pieces of fabric to fit your duvet add 1.25cm/½in seam allowance on three sides and 5cm/2in for a deeper hem at the opening on the top edge. Cut 2 scallop-edged borders 15cm/6in wide to fit two sides and bottom edge of the duvet. Allow 10cm/4in extra for gathering around the corners. Right sides facing sew the scalloped edges together. Turn right side out.

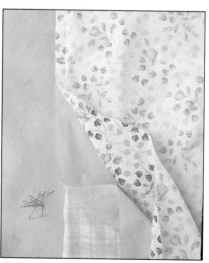

2 Turn under the allowance at the opening to form a double hem, pin then sew in place. Place the right sides of fabric together and sandwich the scalloped edge in between so all the raw edges are even. Pin, then sew in place.

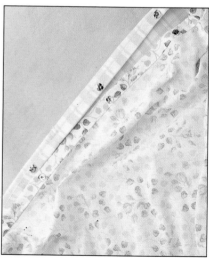

3 Sew large press studs at regular intervals across the opening. Turn the duvet right side out and pull the corners to even out the gathers.

Deckchair

Fabric paint will change the
look of your deckchair. Use
the old cover as a template
and cut new canvas to the
same dimensions.

MATERIALS

pattern paper

canvas

scissors

spray adhesive

coloured spray paint

staple gun/tacks

Hints

Try to find deckchair canvas that is made
to the same width as your deckchair
rather than having to turn over the seam
allowance down the sides.

1 Remove the old cover and cut the new canvas to these dimensions. I have created this tiger print using torn strips of paper, in varying widths, to form the stripes.

2 Use spray adhesive to stick all the strips onto the deckchair canvas. Apply the strips in a diagonal direction from the left and right hand sides to meet at the centre.

3 Spray the paint all over the canvas and paper strips. Leave to dry and then carefully peel off each strip to reveal the tiger print underneath. Secure the canvas to the deckchair frame using staples or tacks.

Rosy Tieback

Even the plainest curtains can look really special if you make this simple, yet very pretty, tieback.

MATERIALS

calico fabric

pattern paper

pencil

iron-on interfacing

piping cord

pins

scissors

PVA

household paint brush

spray paint

thread

Hints

I find that any type of spray paint works on the flowers, even car paint, so don't worry about trying to find spray-on fabric paint.

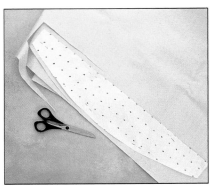

1 Decide on the length of the tieback by measuring around your curtain. Make a template approximately 10cm/4in wide by the length required. Use a pencil to curve the two sides so each measures 4cm/1½in at their ends. Cut two pieces of fabric, adding 1.25cm/½in for turnings.

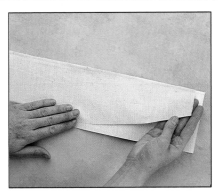

2 Press a piece of heavy weight, iron-on interfacing to one of the tieback shapes. Use only enough pressure on the iron to hold the interfacing in place as this will be trimmed later.

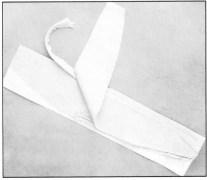

3 Cover enough piping cord to go all the way around the tieback. Raw edges together, pin and stitch this to the stiffened tie-back. Take the second tieback piece, and right sides together, sandwich the piping between, pin then sew along the top edge only.

4 Trim seam allowances on stiffened piece and turn right side out. Cut two pieces of fabric 2.5 x 5cm/1 x 2in. Turn under the raw edges and stitch closed. Fold each in half to form two loops. Place each inside the narrow edges of the tieback and stitch in place.

5 Thin PVA with an equal quantity of water and brush over a piece of cal-ico to stiffen. When dry cut into 15 x 4cm/6 x 1³⁄₄in strips. Fold these in half. At one end fold in the two corners to form a triangular shape.

6 Hold the base of six petals to create a rose. Bind the base with strong thread. Cut out a leaf shape from the stiffened calico. Use red spray paint to colour the roses and green for the leaves. Sew or stick roses and leaves to the tieback.

Dress Curtain & Tails

This dress curtain is designed to loop over a pole in a series of swags and fall at one side with a decorative finish.

MATERIALS

pattern paper

fabric - top

fabric - lining

scissors

pins

string

pencil

needle

thread

Hints

The easiest way to estimate the length of the swagged curtain is to wind a length of string around your pole creating the swags and tails you want to achieve in fabric.

1 Cut your fabric to the length of the string adding 1.25cm/½in all round for turnings. Cut a 90cm/36in square from pattern paper for the tail template. Mark a point 30cm/12in up from the bottom left corner. Mark another point 30cm/12in to the right of this. Draw a line to join the points. Cut out.

2 Position the template on the front of the fabric, on the left hand corner of the narrowest edge. Pin then cut the fabric along the diagonal line, extending it as far as you need to cut the corner off the fabric. Cut a piece of lining fabric to the same shape.

3 Pin, then sew the two pieces of fabric together using the seam allowances. Leave a gap to turn the fabric to the right side, then stitch closed. Press the seams flat. Drape over the pole, pleating the tail into neat folds.

Shower Curtain

There's no reason for you to live with a boring plastic shower curtain when you can have one as beautiful as this.

MATERIALS

organza

plastic

scissors

tissue paper

thread

eyelet punch

metal eyelets

scallop shells

emulsion paint

brush

soft cloth

drill

needle

Hints

Place tissue paper between the organza and the plastic shower curtain to stop the fabric from slipping and to make the sewing easier.

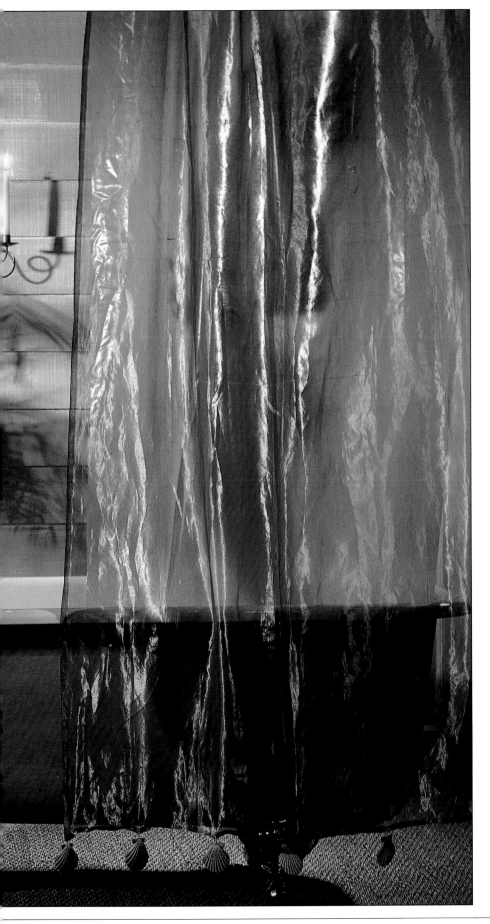

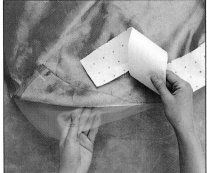

1 Cut the plastic lining and fabric to
the required shower curtain length.
Allow 2.5cm/1in for seams along the
top of the plastic and at top and bottom edges of the organza. Turn in
seams and sew organza to the plastic
on the top edge. Sew the bottom hem.

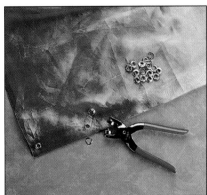

2 Use the punch to form the eyelets
along the top of the curtain. Space
the eyelets at regular intervals making
as many eyelet holes as you have
hooks. Take care to ensure that the
shiny top surface of the eyelet is at the
front of the curtain.

3 Use a paint brush to apply a thin
layer of very watery emulsion paint
to each scallop shell, rub off the excess
paint with a soft cotton cloth. Use a
drill with a narrow bit to create a tiny
hole at the base of each shell. These
holes will enable you to sew each shell
to the shower curtain hem.

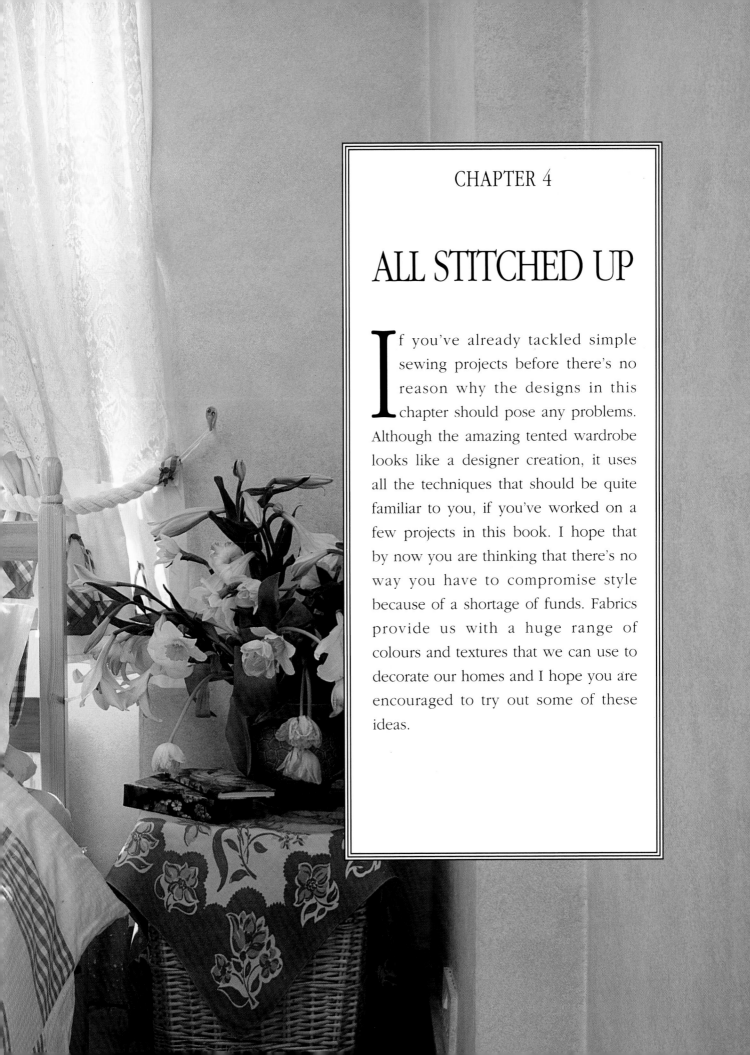

CHAPTER 4

ALL STITCHED UP

If you've already tackled simple sewing projects before there's no reason why the designs in this chapter should pose any problems. Although the amazing tented wardrobe looks like a designer creation, it uses all the techniques that should be quite familiar to you, if you've worked on a few projects in this book. I hope that by now you are thinking that there's no way you have to compromise style because of a shortage of funds. Fabrics provide us with a huge range of colours and textures that we can use to decorate our homes and I hope you are encouraged to try out some of these ideas.

Director's Chair Cover

This lovely slip-over cover transforms the boring old director's chair so it can be used anywhere in the home.

MATERIALS

3 patterned fabrics

pattern paper

scissors

pins

thread

Hints

Make a simple cushion cover from any scraps of fabric that may be left after covering your chair.

1 I have chosen three different fabrics for my director's chair. They are all linked by colour, in this case pale yellow and by design, as all three fabrics have a rose featured in their pattern.

2 Cut pattern paper templates to the dimensions of the director's chair. I find it easier to lay the paper over the chair and cut while in place. When you cut the fabric add 1.25cm/½in all round for seam allowance. Use a zigzag stitch over the raw edges to stop the fabric from fraying.

3 I used: one long piece from the front hem over the seat and backrest to the lower hem; 2 outside pieces; 2 inside pieces; 1 strip over the armrest; 1 strip between the front and back of the backrest. Pin, then sew the pieces together. Turn up the hem, take care when you sew in the corners of the chair. Press each seam open before progressing. Clip the seams where necessary particularly at corners such as these armrests. Turn the cover to the right side and slip over the chair.

Linen Basket

This basket was destined for the dustbin, but a loose bag inside, with several pointed and wired ties around the top edge, means that when the basket is full you twist them together and all your laundry is hidden away inside.

MATERIALS

fabric

pattern paper

piping cord

plastic-coated wire

thread

scissors

emulsion paint

long-haired brush

Hints

Use garden wire with a plastic coating as this will not rust if the fabric gets wet.

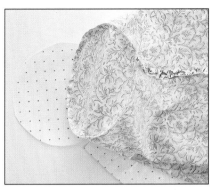

1 Cut a paper template to make the loose bag. Cut one piece for the circular base and another piece for the sides of the basket. Add 1.25cm/½in all round for seam allowances. Cut out then sew together. You may need to make little tucks in the fabric to enable it to fit the circular base.

2 Cut 4cm/1¾in wide strips of fabric to make the casing for the piping cord. Join the strips to form one long length. Lay the cord and garden wire inside the case. Use the zipper foot to stitch closed.

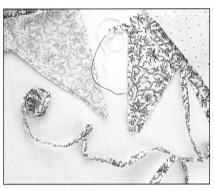

3 Cut template 40cm/16in deep by the circumference of basket. Divide by six and shape each section to resemble petals. Place on double layer of fabric add 1.25cm/½in for turnings and cut out. Right sides facing sew the piping between two petal shapes.

4 Cut a 15cm/6in wide facing from the fabric to fit circumference of basket. Pin, then sew this to bottom edge of petal shapes. If you feel you need to tack the pieces in position before you start to sew on the machine, remember to remove the thread afterwards.

5 Turn the facing to the wrong side of the petal edge. Right sides facing sew the petal edge to the bag. Turn right side out so that the facing covers the top edges of the bag. Hem the facing. I painted my linen basket with bright blue emulsion to match the fabric. Use a long-haired household brush to get the paint into the intricate weave of the wickerwork.

Patchwork Tablecloth

Although not strictly patchwork this tablecloth makes use of two fabrics that are sewn together to make a striking cover for this kitchen table.

MATERIALS

tape measure

fabric – main

fabric – border

scissors

pattern paper

dustbin lid

tassels

needle

thread

pins

Hints

When you have cut all the pattern pieces use a close zigzag to finish all the raw edges to stop the fabric from fraying.

1 Measure the length and width of the table. Cut a piece of fabric to this size and add 1.25cm/½in for the seam allowance. Using a contrasting fabric cut out four border strips, 10cm/4in wide and long enough to edge the cloth. Cut a paper template for the curved sections using a clean dustbin lid to give you a large curve.

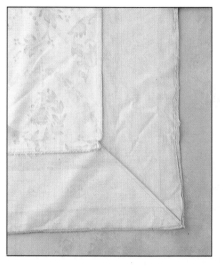

2 Right sides facing pin then sew the border strips to the main fabric. Allow a good overlap at each corner. Sew right into the corner point. Diagonally fold back the strips. Sew along this diagonal line, where the strips meet, to form a mitred corner.

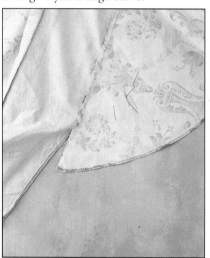

3 Right sides facing sew the curved fabric section to the border fabric. The curves should be placed so they do not quite meet the mitred corners. Turn in and sew around the hem of the curves.

4 Sew a ready-made tassel to each of the four points. Press the cloth with an iron and place it on the table at an angle with the tassels at the centre of the four sides, to show off the softly rounded edges.

Headboard

The beauty of cutting your own headboard is that you can create any shape you like, from a classic arch to something curved and fanciful.

MATERIALS

MDF (medium density fibreboard)

wadding

PVA

fabric – headboard

fabric – border

fabric – piping

piping cord

sewing pins

thread

pencil

ruler

scissors

staple gun

needle

Hints

I have cut a new shape for my headboard using MDF, but you could recover an old headboard simply by removing the old fabric.

1 Position the wadding over the headboard and cut out. Don't worry about cutting this to cover the sides of the headboard, because when you cover it with the fabric the wadding will be pulled over the edges enough to soften them. Use a little PVA to stick the wadding to the front of the headboard.

2 Lay the headboard on a flat surface – the floor is probably the easiest place. Put the fabric over the wadding. Take time to position it carefully and match the pattern with the shape of the headboard. If you are using fabric with an obvious stripe or check make sure this is perfectly square and the lines on the pattern are parallel.

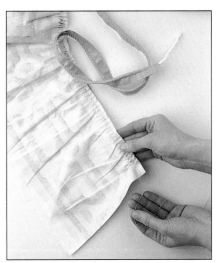

3 Cut a 20cm/8in wide border from the fabric. This should be one and a half times the length of the headboard edge. Sew two lines of stitches on one of the longest edges. Pull these threads to gather the fabric. Using a contrasting fabric cover two lengths of piping cord. Sew to ruched edge of the border allowing more fullness where it bends round a curve.

4 Use a pencil and ruler to mark the position of the border on the headboard about 18cm/7in in from the outer edge. Place the border right side down on the headboard so its seam matches the pencilled line. Pin in place. Turn border over and check its position. Staple in place, or use a curved needle to sew the border securely to the headboard fabric.

5 Pull the remaining edge of the border taut and staple to the back of the headboard. Cover another length of piping cord with a contrasting fabric to fit around this outer edge.

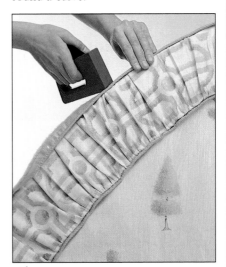

6 Take the second length of covered piping cord and staple or use a curved needle to sew the top edge of the headboard. To finish, roll the piping back slightly with your fingers and using tiny stitches sew down to hide the row of staples or stitches.

Antimacassar

Popular in granny's time, this attractive chair or sofa cover will not look out of place today.

MATERIALS

fabric – antimacassar

fabric – piping

piping cord

self-cover buttons

tape measure

pattern paper

pencil

scissors

pins

tassels

needle

thread

Hints

You can use ready-made tassels for the decoration, but make sure these are not too bulky as they will be uncomfortable to rest your head against.

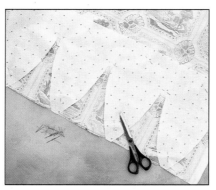

1 Measure the back of your sofa and use this as the width of the antimacassar. The depth is 90cm/36in. Cut out two pieces of fabric to this size. Make a paper template for the shaped bottom edge of your antimacassar. Pin to fabric and cut out allowing 1.25cm/½in all round for turnings.

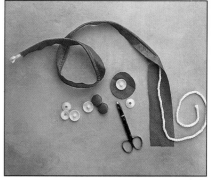

2 Cut 6cm/2½in wide strips in contrasting fabric for piping casing. Join where necessary until you have enough to go all around the edges of the antimacassar. Cut circles in contrasting fabric, slightly larger than the buttons. Pinch the fabric edges together and snap the button base over them.

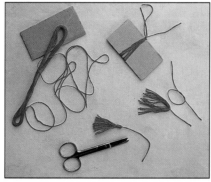

3 To make the tassels, wind embroidery or crochet thread ten times around a piece of card. Remove from card. Wind thread around one end to secure, cut other end to form the tassel shape.

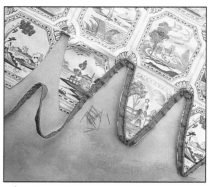

4 Right sides facing, pin the covered piping cord to one piece of the antimacassar. Place the casing so that the corded edge is facing the centre of the fabric. Snip both antimacassar and piping fabric at the corners to ease sewing. Sew in position.

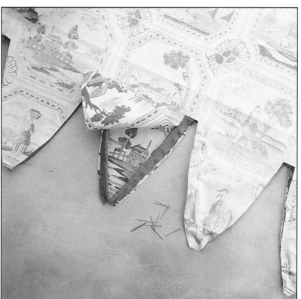

5 Pin the remaining fabric, right sides facing, over the piped fabric. Sew carefully taking care around the corners. Leave a 15cm/6in unstitched gap to allow you to turn the antimacassar right side out, then sew the gap closed. Sew the covered buttons to both sides of the antimacassar to make it reversible. Sew each of the tassels to the base of each pointed triangle for an attractive decoration.

Bathroom Chair

Because this cover is made from soft, absorbent towelling fabric it is ideal for use in a bathroom.

MATERIALS

pattern paper

scissors

pins

towelling

dinner plate

needle

thread

lining fabric

foam padding

bias binding

Hints

Use a long stitch on your sewing machine to stop the towelling from puckering. Zigzag over the edges of each piece before sewing to prevent the towelling from fraying.

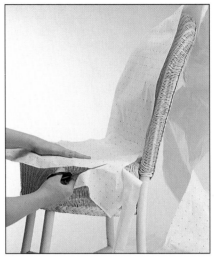

1 Position pattern paper over the chair and cut templates for each of the main pieces. Every chair will be different but generally you will need templates for the seat; from seat to top; from back to the floor and finally a skirt to cover the legs around the sides and front of the chair.

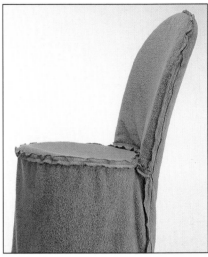

2 Pin each template to one thickness of towelling. Cut around the template adding 1.25cm/½in all round for the seam allowance. Use the chair to pin the pieces together adjusting them to fit. Sew the seams and press open. Position back on the chair to check the fit is correct.

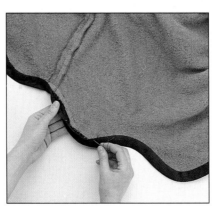

3 Cut a wavy hem all round the bottom edge of the cover. Use a dinner plate to cut the curves accurately, adjusting the measurements to create an even number of curves. Right sides facing, sew bias binding to the hem. Fold over the binding to the back of the cover sandwiching the raw edge in between and hand sew to finish.

4 Using the seat template cut a single piece of lining fabric then cut two pieces of towelling, adding 2.5cm/1in seam allowance. Sew all three pieces together using the lining cloth as a sewing guide. Leave a turning gap. Turn the cover to the right side. Fill with foam, cut to the same size. Sew up the opening.

Roller Blind

This is a simple blind that would look attractive at any window in the home, either on its own or with curtains.

Hints

If you have a window with a recess, position your roller blind against the glazing. If you don't have a recess, mount the blind on the wall above the frame.

Preparation

Cut fabric to fit the window. Thin PVA with an equal quantity of water and apply to fabric to stiffen. Cut a paper template to the width of the blind. Hold a length of string on the paper and twist around until you have a pleasing pattern, draw around the string. Cut out the shaped edge.

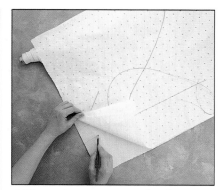

1 The easiest way to transfer the pattern is by using a darning needle to punch holes through the paper template and through the fabric. This will leave a series of small holes or dots in the fabric. Remove the template then simply join up the dots using a soft pencil. Cut along shaped edge.

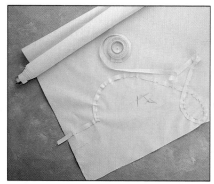

2 Pin the ribbon to the fabric, following the pencil line. Sew the bottom edge of the ribbon in place first. Iron flat then sew the top edge. If you are sewing by hand, use small stitches so the untidy threads are all at the back of the blind.

3 Cut the excess fabric as close as you can to the ribbon edge. The fabric should not fray because it has been stiffened. Attach the cord pull provided in the kit, or a tassel, to enable you to pull the blind up and down. Hang the blind in position.

Austrian Blind Pelmet

An Austrian blind can sometimes look too fussy, but I think this pelmet gives a window with standard headed curtains a softer, more extravagant look.

MATERIALS

fabric	scissors
Austrian blind tape, cord	needle
heading tape	expanding tape measure
thread, trimming	pins, cleat

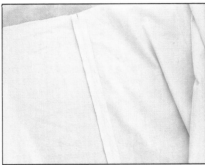

1 Measure the width of your window and cut your fabric to twice this width. The drop of the pelmet will be 50cm/20in. Allow 1.25cm/½in seam allowance all round. Join the widths if necessary with a flat seam.

2 On both side edges fold under fabric by 1.25cm/½in, turn under again and sew. On the wrong side of the fabric mark the positions for the looped tape. The rows of tape should be 30cm/12in apart. Make sure the loops match up horizontally. Pin, tack then sew in place.

3 Turn under a 1.25cm/½in hem along the top edge of the pelmet and pin, tack then sew the heading tape over this. When gathered to the width of the window wind excess cord around a cleat. Thread the cords from bottom to top through the looped tape.

4 The type of fringed edging I have used seals in the raw edges. If your trimming does not have this, turn under a double seam, as for the sides of the pelmet. Mount onto a pelmet track and make final adjustments to the heading tape and pull the cords up until you have the desired effect.

Tented Wardrobe

This wardrobe is created by extending the top shelf and giving it a slightly curved edge. The tented top section will overlap this. Fix a curtain track to the front edge of the shelf.

MATERIALS

cup hook

tape measure

pattern paper

scissors

buckram

fabric

piping cord

contrasting fabric

dinner plate

heading tape

thread

cleat

brass ring

curtain track

curtain hooks

curved shelf

Hints

Screw a hook into the ceiling centrally above the wardrobe. Use a metal tape to measure the height and width of the tent area. Divide the width into equal parts to give you the number of tent sections.

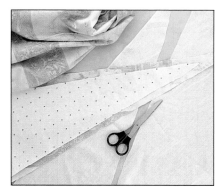

1 Cut a paper template for the tent section. Use to cut out the buckram sections as a backing first and then cut the fabric pieces. Allow 1.25cm/½in all round the fabric for turnings. Sew fabric to buckram with straight stitch close to the edges.

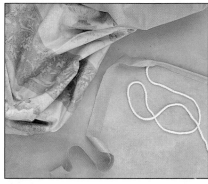

2 Cover the piping cord with contrasting fabric. Use a tape measure to calculate the amount of piping you'll need, remember to include the piping across the bottom of each section.

3 Sew all the tent sections together inserting the piping between each section. Turn the top ends of the piping neatly over to the reverse side where they meet at the pointed top of the tent.

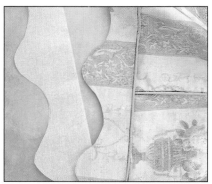

4 Use a dinner plate to make a template for a scalloped border. Cut out in buckram and fabric. To join buckram and fabric, sew around the scalloped edge. Sew the top edge of the border to the tented section again sewing piping cord between.

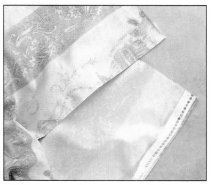

5 Make an unlined curtain for the wardrobe. Measure the drop, and 1½ times the width. Add 2.5cm/1in for the bottom hem and 5cm/2in for the heading. Also add a 2.5cm/1in seam allowance down either side. Cut the curtain and sew down the allowances.

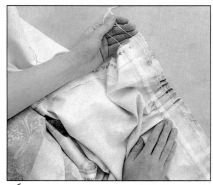

6 Sew the heading tape to the top of the curtain and pull in the gathers, secure to a cleat and use curtain hooks to hang the curtain. Sew a ring to the top of the tented section and hook onto the ceiling. Pull the tented area over the shelf and heading tape.

Dressing Table

The traditional kidney-shaped dressing table has been given a new lease of life with this beautiful fresh floral print.

MATERIALS

tape measure

fabric

curtain heading tape

pattern paper

scissors

pins

lining

interfacing

piping cord

contrasting fabric

needle

thread

pearl drops

Hints

I have scalloped the border of this dressing table for a pretty edge.

1 Make the skirt section first. Measure the depth of the dressing table and one and a half times the track for the width. Allow an extra 15cm/6in for the front overlap and 1.25cm/½in for turnings. Sew the hem and the side seams. Sew the heading tape to cover the raw edges along the top of the skirt.

2 Cut pattern paper to fit the table top plus 1.25cm/½in for turnings. Pin to fabric and cut. Cut a fabric and lining border to fit top, 15cm/6in deep plus turnings. Cut a strip of iron-on interfacing to this size less the seam allowance. Iron to the fabric border.

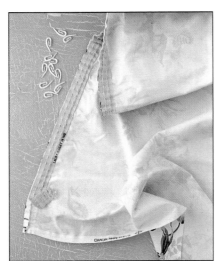

3 Cover piping cord to fit around the table top and border. Right sides facing sandwich the piping cord between border and lining around scalloped edge. Turn right side out. Sandwich more piping between cover top and border, sew together.

4 Cover another length of piping cord to go around the hem of the dressing table skirt. Right sides facing pin then sew in place. Sew on the pearl drops to border edge.

Baby Quilt

This padded quilt is designed to be reversible so the front and back are made in the same way.

MATERIALS

patterned fabric

scissors

needle

thread

pins

wadding

narrow ribbon

Hints

If you want ribbons on both the front and back of the quilt simply turn the quilt over and repeat the procedure making sure the bows are in the same place on each side.

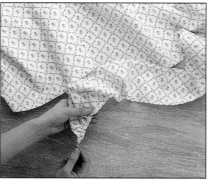

1 Cut two main pieces for the quilt 50cm/20in square and eight border strips 10 x 100cm/4 x 40in. Cut the ends of each strip at an angle of 45 degrees to form a mitre. Hand sew along the top and bottom of each strip to gather.

2 Pin then tack the gathered borders to the four sides of each main piece, adjusting the gathers to fit the corners. Machine sew each mitred corner carefully, then machine sew the border in place, ensuring that the gathers are even. Press the gathers flat.

3 Cut two pieces of wadding to the size of the completed front and back pieces. I have shown you the stitching lines on the wadding, but you could sew the wadding to the fabric following the diagonal lines of the patterned fabric. Stop before you reach the border edge.

4 With the right sides facing and raw edges even, pin the quilted pieces together. Machine sew the seams, remembering to leave a gap on one side, wide enough to turn the quilt through. When the quilt is pulled through hand stitch the gap closed using a needle and matching thread.

5 Thread the narrow ribbon through a needle and push the needle through all the layers of quilted fabric from front to back then to the front again. Cut the ribbon and tie securely using a double knot then a bow.

Appliqué
Tablecloth

You could make this cloth as
a decorative cover that's
replaced with a linen cloth
for meal times.

MATERIALS

tape measure

pattern paper

scissors

black and cream fabric

thread

pins

dressmaker's chalk

dinner plate

piping cord

Hints

If you position your pins at right angles
to the appliqued fabric, you will not
need to remove them as you sew. The
machine will simply ride over them and
your appliqué will be held securely in
place.

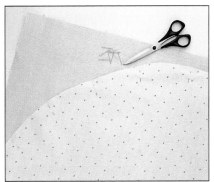

1 Measure the size of your table. To these dimensions add 1.25cm/½in seam allowance all round. Cut a paper template from newspaper or pattern paper. Pin the template to the cream-fabric and cut out.

2 Cut a template or use dressmaker's chalk to mark out the black fabric with irregular, pointed shapes to form the appliqué edge. Remember that if your fabric is an oval shape, like mine, the black fabric should be cut on the cross and the design curved to follow the lines of the tablecloth.

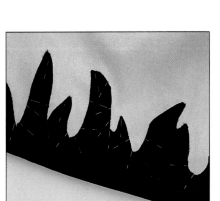

3 Pin the black shapes to the cream fabric. Using a close, zigzag stitch, sew to the table cloth. Where one cut piece of fabric meets another again join with close zigzag stitch. This is the simplest way to cover up the raw edges.

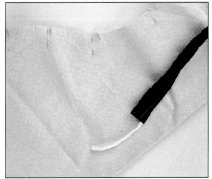

4 Measure the circumference of the table and cut two borders, to this width and 15cm/6in deep. Draw around the curve of a large dinner plate to form the wavy bottom edge. Right sides facing sew the two pieces together. Turn to the right side. Cover piping cord with the black fabric.

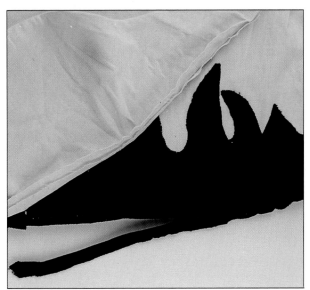

5 Pin the piping cord to the tablecloth along the seam allowance. Sew in place. Right sides facing pin the wavy border to the cloth sandwiching the piping cord between the layers. Sew carefully and trim the seams to remove any bulkiness.

Patchwork Bed Cover

This is patchwork at its simplest and yet it gives a beautiful decorative look to an otherwise plain bed cover.

MATERIALS

fabric -main

fabric – check and plain

pattern paper

thread

scissors

pins

Hints

Use your mattress measurements to cut your fabric. The cover should be measured lengthways, from just under the pillows to the base of the divan and widthways, at the bottom edge of the divan from one side to the other.

Preparation

Cut two rectangles from your main fabric. The rectangle for the top of the cover will be 15cm/6in smaller all round than the bottom. Add a seam allowance of 1.25cm/½in for the three sides and a 7.5cm/3in hem for the top edge. This will be the opening edge.

1 Cut 15cm/6in patchwork squares in both the checked and plain fabrics. Sew these together alternately to form three border strips. Turn under all the raw edges. Pin one strip to the base edge of your top cover. Then pin on the two sides to co-ordinate with this. Sew in place.

2 Use the pattern paper to cut a template 15 x 10 cm/6 x 4in for the trims. Shape as shown. Cut out trims in a double thickness of both checked and plain fabrics. Allow for turnings. Place two trim pieces right sides together and sew around the curved edges. Turn right sides out and press. Repeat with the remaining trim pieces.

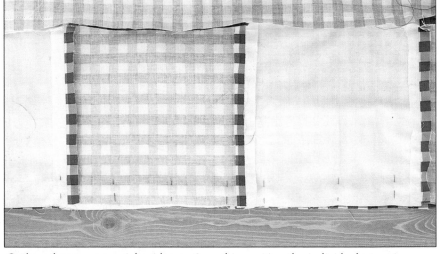

3 Place the top cover right side up. Over this position the individual trim pieces right sides facing, alternating the positions of the blue and white fabric. Pin and stitch. Turn the cover to the right side out and press the trim edge flat. Turn under the allowance for the hem along the top of the cover. Cut a 10cm/4in wide gingham strip to this length. Right sides facing pin then sew one long edge to the top of the cover. Fold over the gingham to cover the seam and hand sew in position.

ACKNOWLEDGEMENTS

The publishers would like to thank the following companies for providing fabrics and accessories for use in our photographs.
Hesse and Company; Laura Ashley; Osborne and Little; Colefax and Fowler; V V Rouleaux; Sandersons; Anna French; The Dormy House; A. B. Woodworking; Swish; Rufflette; Crucial Trading; C.P. Hart; Tintawn; Artisan.

GLOSSARY

British	American
blind	window shade
calico	muslin
concertina fold	accordion pleat
cushion	throw pillow
dress curtain	drapes
dustbin	trash can
emulsion paint	latex paint
festoon blind	balloon shade
frill	ruffle
glazed cotton	polished cotton
kitchen dresser	hutch
muslin	cheesecloth
oversew	overcast